Summary and Analysis of

HIDDEN FIGURES

The American Dream and the Untold Story of the Black Women Mathematicians Who Helped Win the Space Race

Based on the Book
by Margot Lee Shetterly

WORTH BOOKS
SMART SUMMARIES

All rights reserved, including without limitation the right to reproduce this book or any portion thereof in any form or by any means, whether electronic or mechanical, now known or hereinafter invented, without the express written permission of the publisher.

This summary by Worth Books is based on the 2016 hardcover edition of *Hidden Figures* by Margot Lee Shetterly published by William Morrow.

Summary and analysis copyright © 2017 by
Open Road Integrated Media, Inc.

ISBN: 978-1-5040-4665-7

Worth Books
180 Maiden Lane
Suite 8A
New York, NY 10038
www.worthbooks.com

WORTH BOOKS
SMART SUMMARIES

Worth Books is a division of Open Road Integrated Media, Inc.

The summary and analysis in this book are meant to complement your reading experience and bring you closer to a great work of nonfiction. This book is not intended as a substitute for the work that it summarizes and analyzes, and it is not authorized, approved, licensed, or endorsed by the work's author or publisher. Worth Books makes no representations or warranties with respect to the accuracy or completeness of the contents of this book.

Contents

Context	1
Overview	3
Summary	7
Timeline	31
Cast of Characters	35
Direct Quotes and Analysis	39
Trivia	43
What's That Word?	45
Critical Response	49
About Margot Lee Shetterly	51
For Your Information	53
Bibliography	55

Context

Hidden Figures was published in 2016, following years of research by Margot Lee Shetterly. As a Hampton Roads native whose father worked for NASA, Shetterly was already intimately connected to the story of these forgotten American heroes. With *Hidden Figures*, she aimed to break down the historical silos that divide our nation's history and honor the achievements of the African American women in the space program who, despite outrageous obstacles, contributed so much to America's domination of the heavens.

In addition to telling the story of NASA's female mathematicians, *Hidden Figures* recounts America's civil rights movement from the 1940s to the 1970s. It describes the landmark cases such as *Brown v. Board of Education*, the March on Washington, the pay disparities that were systemic and sanctioned between whites and blacks (and especially between women of color and white women), and what it was like for women who sought higher education in mathematics and sciences and who wanted to pursue careers beyond teaching—which were few and far between. The space race—and America's need to beat the Soviet Union during the Cold War era—brought the women in this story, and women in general, to the forefront of math and science careers. Without them, America's journey to the moon, and stars, would not have been the same.

Author's Note on Language Used in the Book

While common parlance would replace the formerly used word "Negroes" with "African Americans," Margot Lee Shetterly has opted to use the following terms throughout the book: "Negro" and "black." Though it may be jarring to modern-day readers, this summary and analysis honors her choice of language.

Overview

In the early 1940s, America was in the thick of World War II. The only hope to fend off the aggressive German forces was to increase America's capabilities in the sky. Langley, Virginia, was home to the leading research facility dedicated to developing aircraft advancements—making planes faster and deadlier than ever. Yet they had a problem with their workforce: they simply couldn't find enough qualified mathematicians to keep their engineers working at top speed. So they started recruiting women. Women were already working industrial jobs as part of the war effort, but this was one of the first occasions in American history that women were sought for technical and skilled positions.

SUMMARY AND ANALYSIS

At first, only white women were considered. But when enough white women couldn't be found, Langley made the decision to also seek out black women to fill these roles.

In a deeply segregated nation where "separate but equal" meant anything but, the Langley Memorial Aeronautical Laboratory would find itself at the epicenter of an entire movement for women, and especially black women, in the fields of engineering and mathematics. To house the new recruits, Langley formed a division on the campus for black female "computers," or women who ran calculations, called the West Area Computers. It was here that the women who ultimately made America's moon landing possible would find their start in the leading technical research facility in the nation—and perhaps, the world.

Dorothy Vaughan was a schoolteacher with a knack for math. She graduated from college and promptly went into teaching, one of the highest paying jobs available to black women in the 1940s. After executive orders were issued in 1941 making it possible for Negroes to apply for civil service jobs, Dorothy spied a notice in her local post office. The advertisement called for women to be "computers" at a place called Langley. She applied, was accepted, and on December 1, 1943, she began working.

Mary Jackson, who held degrees in both mathematics and physical science, started in Dorothy's West Area Computers pool. She would eventually earn a position in the widely acclaimed engineering team of Kazimierz Czarnecki, one of Langley's most revered engineers.

Katherine Goble was from Hampton, Virginia. Twelve years earlier she had left graduate school, where she had been pursuing a degree in mathematics, to raise her first child. After she was accepted into Langley, she began work with Dorothy in the West Area pool in 1953. There, she forged a strong friendship with Eunice Smith, another West Area Computer. Katherine was eventually assigned to the advanced Flight Research Division.

The West Area Computing pool was dissolved in 1958, the same year that NACA (the National Advisory Committee for Aeronautics) became NASA, transforming the group's mission from perfecting airplane flight to perfecting space flight. Dorothy, Katherine, and Mary continued to work at NASA.

On February 20, 1962, John Glenn was the first American to orbit the earth, thanks to calculations he specifically asked Katherine Goble (now Johnson) to check and recheck. Then, in 1969, America put a man on the moon. Through it all, these brave women shaped the future of America's space program, refusing to let the grip of racism and gender bias diminish their talents and potential.

Summary

Chapter One: A Door Opens

In 1943, the Langley Memorial Aeronautical Laboratory was vastly undermanned. To meet staffing demands, women were hired as "computers," mathematicians who did the calculations for the engineers—which was unusual even in post–World War II America, where men still dominated scientific and technical jobs.

Langley sent female scouts up and down the East Coast in search of white women with the skill and talent necessary to support their growing demand. But they were still unable to fill the empty positions. Thanks to two executive orders issued by President Roosevelt in 1941, Negro Americans had become eligible for civil service jobs through the desegregation of the defense industry. It's here that Langley found the talent it needed.

As Langley began to hire qualified Negro female candidates, they built accommodations to house this new pool of workers. While the folks at Langley were accustomed to seeing Negroes in unskilled roles, such as groundskeepers and janitors, it was something entirely different to consider black women as professional peers. Little did they know, these women would help land the United States on the moon.

The Langley Laboratory played a key role in US aeronautics history. The testing group was made up of civilians on the Langley government installation called the National Advisory Committee for Aeronautics (NACA), which was the predecessor to NASA (National Aeronautics and Space Administration).

Chapter Two: Mobilization

Dorothy Vaughan was born in 1910. As her mother passed away when Dorothy was only two years old, her stepmother raised Dorothy as her own and encouraged her to excel academically. This focus on education served Dorothy well, earning her a full scholarship to Wilberforce University, the nation's oldest private Negro college. Although encouraged by her family and professors to pursue her graduate degree in mathematics at Howard University, the cost seemed extravagant in the wake of the Great Depression. After graduation she opted instead to join the workforce and support her family, becoming a teacher in 1929.

During World War II, as fighting raged overseas, there was always an influx of new civil service job postings at the local post office. In 1943, Dorothy caught sight of two such postings: one for a position in the Camp Pickett laundry center, where she could earn extra money during the summer; and another for a job at an aeronautical laboratory seeking people with a mathematical background. Dorothy completed both job applications, fairly sure she'd be hired in the laundry, but excited about the prospect of working at a well-paying job at the lab.

Chapter Three: Past Is Prologue

Dorothy Vaughan was a fixture in the Farmville school system, holding her students to exceedingly high standards. As expected, during the summer she worked in the laundry at Camp Pickett, earning forty cents per hour.

Later that year, Dorothy finally received a letter hiring her for the position of Mathematician, Grade P-1, at Langley. Her salary would be $2,000 per year and her job was guaranteed through the end of the war. This was $850 more per year than she earned in her teaching position, and an opportunity she couldn't pass up. Her family—husband and four children—remained in Farmville while she boarded a Greyhound bus bound for Newport News, where she found a room to rent.

Chapter Four: The Double V

When Dorothy moved to Virginia, the towns and cities surrounding the harbor of Hampton Roads—Newport News, Hampton, Portsmouth, Norfolk, and Virginia Beach—were bustling. The population had grown considerably during the war. Dorothy took the bus from her home in Newport News to and from work, facing daily the segregation that dictated where blacks could sit, enter, and exit the bus. While segregation and "separate but equal" laws were promoted as reducing racial tension, they seemed to do anything but. Black schools were ignored and left in ill repair, restaurants refused to serve black customers, and it was no rare instance when uniformed black soldiers were seen as overstepping their bounds, though they were in service to their country just like their white counterparts. Dorothy had been lucky enough to find one of the few quality career opportunities open to Negroes.

Dorothy boarded one of these segregated buses and started work at the Langley Memorial Aeronautical Laboratory on December 1, 1943. It was the same day that the leaders of Great Britain, the United States, and Russia wrapped up their conference in Tehran, where they planned the invasion of Normandy in 1944—D-Day.

Chapter Five: Manifest Destiny

On her first day at Langley, Dorothy Vaughan filled out the paperwork, received a blue badge with her photo on it, and promptly went to work in what would be referred to as the West Area. The West Area was home to such testing equipment as the massive Sixteen-foot High-Speed Tunnel—used for aeronautics testing—and a wide variety of other equipment hidden in buildings and rooms throughout the west campus.

Employees at Langley were segregated. The West Area (or West End) "computers" were all black women. Their white counterparts worked in the East Area computing pool. Margery Hannah was the West Computing's section head.

The white employees at Langley who made efforts to include or even view the campus's black employees as equal were met with disdain. In spite of this, Margery Hannah, a white woman and head of the West Area Computing pool, often invited black women over to her apartment for social functions, and one of the center's leading (white) male engineers was taken into police custody after speaking out about seeing a black man mistreated by the police.

While having to endure these unfair circumstances and social norms, the black computers of the West Area were dedicated to proving that they weren't just something to be kept separate. Rather, they would prove that they were equal to, if not better than, many of their white peers.

Chapter Six: War Birds

The Tuskegee Airmen, a group of African American military pilots, won critical adoration from Americans of all colors, disproving the long-held conviction by whites that Negroes were less competent and ill-suited for the rigors of war. Meanwhile, Dorothy Vaughan was perfecting the four-blade, propeller-powered Mustang aircraft, which was to be the fastest plane designed by Americans yet.

To drive air superiority forward, Langley invested in crash courses for the new West Area computers. At the end of the day, the ladies of the West End would dive into courses on aerodynamics, weekly labs in the center's wind tunnels, and take on homework assignments above and beyond their normal workload.

By studying the way wind flowed over a prototypical wing or fuselage, engineers could build faster aircraft piece by piece. No other laboratory could compete with Langley's wind tunnel research and their engineering talent. Assisting the engineers were the computers like Dorothy, who computed, re-computed, and checked one another's work. Frequently, Dorothy and her coworkers handled a small part of larger computations, oftentimes without specific knowledge of what the calculations were for.

When the United States dropped the atomic bomb on Japan, Langley said that everyone at every level should feel they played a role in winning the war.

Chapter Seven: The Duration

Dorothy frequently made the trip back and forth from Hampton Road to Farmville to visit her family. Despite not knowing how long her job at Langley would last, since the war was coming to an end, she decided to make Hampton Road her permanent home in mid-1944. In the fall of 1944, her four children moved to nearby Newsome with her, leaving her husband to his varied schedule seeking out seasonal hotel-related jobs. He visited when he could and stayed close to Farmville to care for his elderly mother.

When V-J Day came in August of 1945, there was much celebration and a comparable amount of anxiety. There were massive layoffs for many who had been employed to further the war effort. Fifteen hun-

SUMMARY AND ANALYSIS

dred Newport News shipyard workers were let go, and more than two million women workers received their walking papers by the end of August 1945.

During this postwar time, Dorothy grew close to Miriam Mann, a fellow West Area Computer. They attended social functions together outside of work and their families spent time together.

Chapter Eight: Those Who Move Forward

As long as Katherine Goble (neé Coleman) could remember, she loved to count. Her father, skilled in mathematics himself, moved his family 125 miles so she and her siblings could study at West Virginia State Institute. Her father would go to work as a bellman at the Greenbrier Hotel, where he would later meet Dorothy Vaughan's husband.

An exceptional student, Katherine started college at West Virginia State College at age fifteen in 1933. She devoured her studies, causing Dr. William Waldron Schieffelin Claytor, one of the college's leading mathematicians, to handcraft advanced courses for her. It was his recommendation that she begin to prepare for a career in research mathematics. After she graduated, she went to work as a teacher in Marion, Virginia.

There, she met Jimmy Goble and they got married. She taught for two years and then left the profession to stay home. But when her husband fell ill, she took up teaching again to make up for his lost income.

In early 1940, she found Dr. Davis, the president of West Virginia State College, waiting outside her classroom door. He had selected her as one of the first graduate students to integrate West Virginia University's graduate program and she enrolled in the school's 1940 summer session. At the end of her first summer of graduate school, however, she and her husband found out they were expecting their first child, so Katherine left graduate school to be a full-time wife and mother.

Chapter Nine: Breaking Barriers

Dorothy Vaughan took leave from Langley to raise two more children, never questioning that she would return to the work she loved.

Though the government was paring down staff postwar, talented computers remained valuable resources. After Dorothy returned to work, she became a shift supervisor; in 1946, she was finally made a permanent employee.

During this time, Langley employees were sent off to the Mojave Desert to create the Dryden High-Speed Flight Research Center. When Chuck Yeager broke the sound barrier at that facility on October 14, 1947, it was a female computer who was in charge of analyzing the data from his plane.

SUMMARY AND ANALYSIS

The East Computers was disbanded in 1947, and its members took on other projects. The remaining assignments were sent to West Computing. But soon, one of West Computing's own would get her shot at a specialized job. Dorothy Hoover, a talented mathematician, crossed over into a specialty group in 1946, working directly for one of Langley's leading white male engineers named R. T. Jones.

West Computing held a heavy workload, hiring more women of color to keep up with demand. Dorothy Vaughan continued to excel, and when the Head of the West Computing Section fell gravely ill in 1949, Dorothy was tapped to take the lead in the interim. She wouldn't be made the section's permanent head until 1951.

Chapter Ten: Home by the Sea

Mary Jackson grew up surrounded by people who challenged her to succeed. After graduating from Phenix High School with honors in 1938, she enrolled at the Hampton Institute for college, pursuing a double major in mathematics and physical science. After graduation, she became a teacher, but had to leave her position to care for her ailing father. At home in Hampton Roads, she took a job at the King Street USO as a secretary and bookkeeper.

While at the USO, she met her husband, Levi Jackson, and in 1946, they had a son, Levi Jr. In her spare time, she led the local Girl Scout troop, putting her charges through real-world rigors instead of giving them badges for projects that were less meaningful.

Feeling threatened by the rise of communism during the Cold War, the United States was looking for allies—and was becoming increasingly worried about its reputation. Foreign leaders were horrified by the treatment of blacks in America. The issue of segregation was becoming one of great concern, prompting a 1951 memo within NACA that discussed whether or not the West Area Computers was a segregated unit.

When Levi Jr. was four, Jackson returned to work and applied for civil service positions as both a clerk at Fort Monroe and a computer at Langley. Langley's West Area Computers was one of the few places where Negroes could find quality jobs, so it was always news when Dorothy Vaughan was hiring. After a mere three months on the job at Fort Monroe, she was offered a position in Dorothy Vaughan's West Computing.

Chapter Eleven: The Area Rule

Dorothy continued to dispatch Negro computers to the specialty groups throughout Langley. Yet while the black women of West Computing intermingled with the whites of other professional groups, there was no doubt that segregation remained. African Americans

SUMMARY AND ANALYSIS

still had their own bathrooms, and they were not getting promoted at the same rate as the white employees. Race interrupted the equity of professional lives.

While on assignment to another section in the building, Mary Jackson casually asked her female coworkers where the restroom was. They giggled, because how would these white women know where a colored girl's bathroom was? Incensed and embarrassed, Mary made the journey back to the West End and happened upon Kazimierz Czarnecki, one of Langley's brightest engineers. In the conversation that ensued, Mary told him about what had just occurred, and on the spot, Czarnecki invited Mary to work on his team. He didn't yet know about her academic prowess, but he'd soon learn.

Chapter Twelve: Serendipity

Twelve years after she left graduate school, Katherine Goble's brother-in-law urged her and her family to move to Hampton, Virginia, where his family lived. He offered to help her husband, Jimmy, get a job painting at Langley, and urged Katherine to join Langley as a mathematician. The Gobles decided to take the chance and move to Hampton.

For the next year, Katherine's family settled into their new home. Jimmy worked as a painter and their children attended school with the black middle class.

In 1952, Langley approved Katherine's application to become a computer, and in 1953, she began work. Eunice Smith, another West Computing employee, lived down the street, and the two began a carpooling routine that would span thirty years.

A mere two weeks after arriving, Dorothy Vaughan assigned Katherine to the Flight Research Division, launching her deep into the specialty of fighter planes.

Chapter Thirteen: Turbulence

After Katherine had spent six months in the employ of the Flight Research Division, Dorothy Vaughan wanted her to either return to West Computing or compel the division to hire and promote her, along with the requisite raise. As her talents had become quickly apparent, she was made permanent.

Her research in the Flight Research Division brought to light new information that would save lives and change flight patterns. Following the analysis of a propeller plane crash, she discovered that planes leave wakes, and with them, the potential of disrupting other planes crossing those paths within a certain time period.

Katherine's family bought a lot and built a home in a new housing development, furthering their climb through the black middle class. But in 1955, her husband fell ill with an inoperable tumor. He

SUMMARY AND ANALYSIS

passed away in late 1956. With family, friends, and colleagues by her side, Katherine returned to work in January of 1957—ready to begin an unexpected second act of her life.

Chapter Fourteen: Angle of Attack

1947 began with Langley buying its first electronic calculator from Bell Telephone Laboratories. What would take the most talented computers a month to calculate now took a matter of hours. The whole building shook when the machine was running.

In the mid-1950s, the center bought an IBM 604, then an IBM 650, dedicating itself to the mechanical age. Despite the machines' speed, they were prone to errors, and human computers checked and rechecked the work of these mechanical monsters. Dorothy Vaughan saw the future in these machines and quickly set to work learning to program them.

Outside the laboratory, the idea of integration was catching fire. Because of the deplorable conditions at Farmville's Negro public school, the students staged a walkout, demanding equal standards to the local white school. Farmville was the town that Dorothy had left behind in the 1940s, and her nieces and nephews were among the strikers. Their demonstration led to the landmark 1954 decision, *Brown v. Board of Education,* which banned segregation in all

US public schools. However, Virginia Senator Harry Byrd refused to accept the *Brown v. Board of Education* decision, holding that integration just wouldn't be welcomed in the South.

After-hours courses were offered through the Hampton Institute, where many of the West Area Computers had attended college. However, the advanced classes were held at Hampton High School through the University of Virginia's extension school. Hampton High School, thanks to Senator Byrd, was off-limits to Negroes.

This meant Mary Jackson would have to get special permission to enter the school grounds from the City of Hampton in order to advance her career on Kazimierz Czarnecki's team. She went to the school district, hat in hand, to request the dispensation to further her education and help her country. Her request was granted.

Chapter Fifteen: Young, Gifted, and Black

Christine Mann (Darden) graduated from arguably the best Negro high school in the country, the Allen School in North Carolina. While many graduates went on to prestigious schools like Smith and Vassar, Christine opted to attend Hampton Institute and was awarded a scholarship by the United Negro College Fund.

Meanwhile, the world watched as nine black teens

SUMMARY AND ANALYSIS

were the first to integrate a white public school in Little Rock, Arkansas. Despite the state's best attempts to halt the integration, the black students, escorted by US troops, entered the school, books clutched to their chests, and struggled to maintain their composure as white students threw garbage and yelled racial slurs at them. In the Cold War competition for allies, the Little Rock crisis was an embarrassment, undermining the legitimacy of American democracy. But in October 1957, Little Rock was quickly forgotten on the world stage: Russia launched *Sputnik*, the first satellite, into space. The dawn of the space age and the race to the stars had officially begun.

Chapter Sixteen: What a Difference a Day Makes

Americans were shocked by the Soviets reaching space. Was it a sign that America was no longer the global political leader? Determined to best their Cold War rival, the United States government committed the nation full force into outachieving the Soviets.

The NACA women were hard at work. Katherine Goble was at the heart of the research group that would lead America's space program. Meanwhile, Dorothy Vaughan watched the racial landscape transform inside NACA. More black women now worked directly for specialty engineering groups than in her pool of computers. However, her group remained all black and was the only stand-alone group of women at Langley. In October of 1958, NACA became NASA, combining Langley, multiple research operations, and the Jet Propulsion Laboratory. Where NACA had been quiet about the work they performed, NASA would be quite public.

West Area Computing was dissolved in 1958, officially ending segregation at Langley. Dorothy had spent fifteen years in West Area Computing, seven of those as its leader. The ending was bittersweet; West Area had defined and embraced her talents, but its ending was also an opportunity for a fresh start.

Chapter Seventeen: Outer Space

When President Eisenhower's administration put out the fifteen-page document "Introduction to Outer Space" in 1958, Katherine Goble and her colleagues read it attentively. The expectation of NASA's engineers was that they would get America to space as quickly as possible.

The NASA engineers and computers were experts in making the world's best planes, but spacecraft was an entirely new world. With the best and brightest engineers leading classes on a regular basis, Katherine and her colleagues quickly began educating

SUMMARY AND ANALYSIS

themselves on everything that had to do with space technology. Katherine Goble was soon tasked with preparing all of the data and charts for these educational lectures. However, she was never invited to them. When she asked why, she was simply told that "girls don't go to meetings."

She remained persistent, unwilling to let gender be an obstacle. In 1958, she was finally allowed to attend the meetings for the editorial board that would soon become the Aerospace Mechanics Division of NASA. Despite the successes of African Americans inside and outside Langley, black women still had to combat racism and sexism daily in order to do their jobs well.

Chapter Eighteen: With All Deliberate Speed

At the end of the 1950s, the American space program was in sorry shape compared to Russia's. Fears of Russian domination of the skies were worsened by the threat of communism. But America had every intention of becoming the world's leader in space travel. NASA's Space Task Group set up shop at Langley and immediately gave a name to the first manned space mission: Project Mercury.

Yet for all NASA was contributing to America, the country itself was still struggling with the issue of race. Why, many people wondered, was the United

States trying to hard to dominate the skies when there were so many unsolved issues on earth?

Norfolk schools were shut down to protest forced integration, leaving 10,000 students at home instead of in the classroom. In Hampton, Dorothy Vaughan, Mary Jackson, and Katherine Goble continued to push their children to excel in their segregated schools.

Katherine Goble found a bright light during these troubled times. She met James (Jim) Johnson, a military veteran and active reservist. The two would marry at the end of 1959.

NASA helped make another match during this time—between their soon-to-be-built spacecraft and the men who powered them: astronauts. The Mercury Seven were presented to the world in 1959, giving faces to America's space-bound ambitions. Now what remained was figuring out how to get these men into orbit.

Katherine became a trajectories expert, put in charge of the path the Mercury spacecraft would take from the moment it launched until the moment it splashed back into the Atlantic.

Chapter Nineteen: Model Behavior

Mary Jackson devoted herself to giving back to her community and supporting other black women in their quests to climb higher in their careers. She helped her son become the first ever black child to win the Virginia Peninsula Soap Box Derby. By showing him how to use engineering knowledge—what she did every day at work—to build a faster than ever soap box car, she inspired him to aim high, too. When Levi Jr. won the race, he was asked what he wanted to be when he grew up: "An engineer, like my mother."

Chapter Twenty: Degrees of Freedom

In 1960, as NASA continued with the Mercury mission, racial protests were organizing across the country. From sit-ins to other demonstrations, America was a hotbed for civil rights activists seeking the recognized equality of all persons, regardless of color.

Langley moved in the opposite direction of Virginia, placing black women in the most technical engineering teams of the space age. Dorothy Vaughan reinvented herself as a computer programmer, learning FORTRAN, which would translate engineering equations into a language a computer could understand for speed calculations. What she used to assign to one of the girls in her computing pool, she now punched into cards that were fed into a machine.

1960 was the year that NASA was finally ready to launch a manned space capsule into orbit. Multiple communication points spread across eighteen locations would track the Mercury capsule throughout its launch, orbit, and return to sea. Yet while NASA was still busy making calculations, Yuri Gagarin became the first man to orbit the earth in a Russian craft on April 12, 1961. America achieved a manned suborbital flight, one that didn't leave the Earth's atmosphere, later in 1961. But it wasn't enough. President John F. Kennedy announced his plans to put a man on the moon—which was a surprise to NASA, especially since they were just getting a man safely into suborbital flight.

NASA, per the government's decision and the hefty influence of Vice President Lyndon B. Johnson, would be moved to Houston, Texas, taking with it the majority of its talented engineers.

Katherine Johnson, however, would not make the transfer and opted to stay in Hampton Roads for her family.

Chapter Twenty-One: Out of the Past, Into the Future

Astronaut John Glenn would be the man aboard NASA's first orbital flight. John Glenn had one request before he was to take flight: he wanted humans, not computers, to check the work the computers had done on the calculations for his flight. Katherine Johnson turned out to be that human. "Get the girl to check the numbers," said Glenn. The numbers, she said, were good.

SUMMARY AND ANALYSIS

On February 20, 1962, John Glenn set off for orbit on the end of an Atlas rocket from Cape Canaveral with 135 million eyes watching on live television. Glenn returned to earth safely, though forty miles off course due to the additional weight on the capsule from the rocket pack that remained attached. Thirty thousand people turned out for the parade in Hampton Roads to celebrate Glenn, NASA, and America's space achievements.

Chapter Twenty-Two: America Is for Everybody

1963 saw the March on Washington with more than 300,000 people gathering in the nation's capital. Dr. Martin Luther King Jr. gave his famous "I Have a Dream" speech off the cuff and without notecards. Dorothy Vaughan would celebrate her twentieth year in service to the government that same year, earning a ruby-appointed lapel pin in commemoration.

Langley became home to many new black engineers in the mid- to late-1960s, with new recruits finding a mentor in the ever-welcoming Mary Jackson.

Christine Darden (neé Mann) joined Langley following her graduation from Virginia State University. She met Katherine Johnson at church and the two quickly became friends, though they would never work together at Langley. Eunice Smith joined the circle. They socialized and ultimately grieved together when a fire in the Apollo 1 capsule, while still on the launchpad in Cape Canaveral, cost astronauts Gus Grissom, Ed White, and Roger Chaffee their lives.

Katherine continued to help NASA, and America, work its way to the moon. She figured out a complicated two-stage system that would get the astronauts into orbit, to the moon, onto its surface, and back. She'd see soon if her work was good enough.

Chapter Twenty-Three: To Boldly Go

On a Saturday night in July 1969, 600 million people watched and waited to see if man would truly make it to the moon. Katherine Johnson tuned in anxiously from a resort in the Poconos where she was attending her Alpha Kappa Alpha sorority conference. Who would have thought that she'd be part of the team that had been key to the Saturn V rocket taking flight four days prior with the Apollo 11 craft affixed?

The Civil Rights Act of 1964 and Voting Rights Act of 1965 made major headway for black Americans, followed by the Housing Rights Act of 1968, but it would be years before their effects could be more fully realized for citizens of color in America. America had managed to put a man on the moon, while black men and women still couldn't enter certain bathrooms.

SUMMARY AND ANALYSIS

Oddly enough, the television show Star Trek would put African Americans on even playing ground in space. Lieutenant Uhura, played by Nichelle Nichols, showed a black woman in charge on a futuristic craft exploring bold new frontiers. She had considered quitting the show, but was implored by Martin Luther King Jr. himself to stay. "We are there because you are there," he said.

At 10:38 p.m. on July 21, 1969, Neil Armstrong set foot on the moon's surface and Katherine Johnson felt the ripple effect of all the women before her. Generations had come together to make it possible for her to be a part of this historic moment.

Timeline

December 1, 1943: Tehran Conference. Dorothy Vaughan begins work at Langley.

1944: Dorothy Vaughan makes Hampton her legal home.

August 15, 1945: V-J Day.

1946: Dorothy Vaughan is made a permanent Langley employee.

1947: Dryden High-Speed Research Center opens.

October 14, 1947: Chuck Yeager breaks the sound

SUMMARY AND ANALYSIS

barrier while flying over the Mojave Desert.

1951: Mary Jackson begins at West Computing.

1953: Katherine Goble joins Dorothy Johnson's West Computing division.

1958: NACA becomes NASA.

1958: President Eisenhower's administration publishes "Introduction to Outer Space."

May 5, 1958: West Area Computing Unit is dissolved.

January 1958: US Army Jet Propulsion Laboratory successfully orbits *Explorer 1* satellite.

1959: Twenty thousand Langley locals visit NASA's first open house.

1959: NASA introduces the Mercury Seven astronauts to the world.

April 12, 1961: Soviet cosmonaut Yuri Gagarin becomes the first man to orbit the earth. President Kennedy announces executive order 10925, which instituted affirmative action in government and civil service jobs.

February 20, 1962: John Glenn becomes the first American astronaut to orbit the Earth.

August 28, 1963: Historic March on Washington.

July 21, 1969: Neil Armstrong is the first man to set foot on the moon.

Cast of Characters

Christine Darden: Christine spent several years at NASA as a data analyst after graduating from Hampton Institute. Instilled with a love for arithmatic at an early age, her proficiency in math is what earned her a spot on Dorothy Vaughan's West Area Computers. She yearned to be an engineer, constantly asking why men were promoted ahead of her. Christine was finally promoted and began her engineering career at Langley in sonic boom research. She would ultimately spend more than forty years at NASA.

Mary Jackson: Mary was an ambitious and highly

SUMMARY AND ANALYSIS

talented black mathematician who would work with some of the most prestigious research groups at Langley. A native of Hampton, Virginia, Mary grew up surrounded by some of the brightest black minds who contributed to the space race. She began her career as a school teacher and ultimately joined NACA, which would become NASA. After more than thirty years of service as a computer and then engineer, she would make a career shift so she would be in a position to help others. While not mentioned in the book, Mary stepped down from her engineering position and transitioned to an administrative position as an Equal Opportunity Specialist. In this role, she shaped hiring policy and decisions and made a lasting impact on the lives of minorities and women throughout NASA.

Katherine (Goble) Johnson: A black research mathematician who would rise through her career ranks at NACA, then NASA, through the thick of the civil rights movement. Katherine accelerated through schooling, beginning high school at age ten as a math prodigy and completing college by age eighteen. She began her career as a teacher and left teaching to raise her three daughters with James Goble. In 1953, she returned to work at NACA as a computer. Her husband became ill and passed away in 1956. She would later marry Lt. Colonel James A. Johnson in 1959. A tenacious and not-to-be-deterred woman, she rose in

NASA's ranks and would become the girl who John Glenn asked to check the calculations for his historic orbit of the earth. Katherine Johnson would go on to be awarded the highest civilian honor in the United States in 2015—the Presidential Medal of Freedom.

Miriam Mann: Miriam was Christine Darden's mother and fellow West Area Computer with Dorothy Vaughan. Hailing from Georgia, Miriam pursued studies in chemistry and math as a college student. She taught for a few years before marrying William Mann. They had three children together, including Christine, and made the move to Hampton, Virginia, when William accepted a teaching position at the Hampton Institute. Miriam would join the West Area Computers in 1943. Miriam was the one who wasn't keen on the "coloreds" sign on the single table at the back of the Langley cafeteria. She was known to steal the signs, slipping them into her purse, to remove the unnecessary label from their dining area. While another sign would materialize a few days later, the sign did disappear one day and never returned.

Dorothy Vaughan: A college graduate and math teacher who would eventually manage Langley's famed West Area Computers. Dorothy was trained as a teacher and was an educational leader in Farmville, Virginia when she applied to be a computer at

SUMMARY AND ANALYSIS

Langley. She was married with four children when she made the move to Hampton Roads to accept her position as a computer at Langley. She and her family shared a commuting relationship for her first year at Langley. After a year, she moved her children up to Hampton and they began school there. Her husband would continue to pursue seasonal work in various places on the East Coast, which put strain on their relationship. Thankfully, her friends at Langley and in Hampton lent a family-like structure to her and her children in her husband's absence.

Direct Quotes and Analysis

"If disaster did befall John Glenn, one secret military document proposed blaming it on Cubans, using it as an excuse to overthrow Fidel Castro."

The space race took place during the Cold War, one of the most difficult times in US history. Both the Soviet Union and the United States were building nuclear programs, and America faced threats from both the Soviets and Cuba.

"The brainy fellas who controlled the computers trusted their computer, Katherine Johnson."

Since computers were such new technology, their accuracy wasn't trusted the way it would be today. John Glenn

SUMMARY AND ANALYSIS

asked for a human computer to check the mechanical computer's calculations prior to his orbital trip. That computer turned out to be Katherine Johnson.

"Tell me where you want the man to land, and I'll tell you where to send him up."

Katherine Johnson's quote to her team of engineers as they were calculating where a craft would have to launch from to land where they wanted it to land. She had become an expert in calculating trajectories.

"The only places on Earth known not to provide free public education are Communist China, North Vietnam, Sarawak, Singapore, British Honduras—and Prince Edward County, Virginia."

United States Attorney General Robert Kennedy, 1963, in response to Prince Edward County, Virginia's refusal to integrate its public schools. Virginia fought hard against the Supreme Court decision that ended segregation in public schools, *Brown v. Board of Education*, and for that reason, desegregation in public schools came much later there than in many other states and counties. This was in stark contrast to the environment at Langley, where African Americans worked alongside whites.

"First in space means first, period. Second in space is second in everything."

Senate Majority Leader Lyndon B. Johnson on the Soviets' achieving space flight prior to the United States. The Soviet Union and the United States were locked in a battle for supremacy on all fronts during the Cold War, and that included the space race.

"Sputnik, some experts declared, was nothing less than a technological Pearl Harbor."

In the context of Americans not having been the first to achieve space, Americans found their world standing in a fragile position as the Soviets took charge in the space race. Just as the Japanese attack on Pearl Harbor had shattered the illusion that the United States could remain safe during World War II, the *Sputnik* launch was a sign that the United States might not be safe on earth—that the enemy was more advanced, and therefore more dangerous.

"Working as a research mathematician at Langley was a very, very good black job—and it was also a very, very good female job."

Dorothy Vaughan faced a choice: Stay in Langley for

SUMMARY AND ANALYSIS

her NACA job, or return to Farmville to teach. She recognized the fact that women—and especially black women—rarely, if ever, were given opportunities to hold jobs such as the one she had at NACA.

Trivia

1. Laundry workers at military installments in 1943 earned forty cents per hour. It was considered a windfall, especially for Negro women with few employment options aside from teaching.

2. In 1945, five out of ten people in southeastern Virginia worked for the government, either directly or indirectly.

3. Published female scientists were rare, but between 1941 and 1945, Doris Cohen published nine reports—five as sole author and four co-authored. This volume was impressive, as many men hadn't been so frequently published.

4. The NAACP (National Association for the Advance-

SUMMARY AND ANALYSIS

ment of Colored People) was a target of the famed anticommunist senator Joseph McCarthy.

5. The US government frequently revoked the passports of those who were outspoken about the rights of blacks during the McCarthy era, such as W. E. B. Du Bois.

6. Black people seeking to purchase homes were frequently denied federally insured bank loans. Thus, North Carolina Mutual Life Insurance Company, a black-owned business, became the largest insurer of Negro mortgages.

7. In 1959, President Eisenhower had a secret bunker built beneath the Greenbrier Hotel in White Sulphur Springs, West Virginia. It would serve as a fallout shelter for heads of state in an emergency and was stocked with a wide array of delicacies, such as champagne and steaks. The bunker was maintained until 1992, when an exposé published by the *Washington Post* blew its cover.

8. Katherine Johnson, a black female engineer, was asked for personally by John Glenn to confirm by hand the computer calculations for his historic Mercury orbit flight.

What's That Word?

Brown v. Board of Education: The historic Supreme Court decision that ruled separate did not mean equal, overturning *Plessy v. Ferguson* which created "separate but equal" accommodations in the United States. A major marker in the civil rights movement that paved the way for desegregation.

Computer: The term used to describe the women who manually ran numbers and equations long before modern-day digital computers were ever put into use. Katherine Johnson, Dorothy Vaughan, Mary Jackson,

SUMMARY AND ANALYSIS

Miriam Mann, and Christine Darden all began as computers at NACA in Langley.

Dryden High-Speed Research Center: A facility established in the Mojave Desert to deal with flight challenges, primarily aircraft design and the pursuit of faster speeds. It was here that Chuck Yeager broke the sound barrier in 1947.

FORTRAN: The computer language that Dorothy Vaughan would take classes in so she could learn to run the new mechanical computers that were brought into Langley. FORTRAN was derived from "formula translation."

NAACP: National Association for the Advancement of Colored People. The NAACP was a leading force throughout the civil rights movement, especially through their Legal Defense Fund. Dorothy Vaughan would be active in the NAACP for the majority of her life.

NACA: National Advisory Committee for Aeronautics, based in Langley, Virginia. This was home to most of the major advancements in aircraft technology during World War II and eventually merged with other government organizations to become NASA. It was home to the historic West Area Computers division.

NASA: National Aeronautics and Space Administration, formed by the government in 1958 to lead the United States' efforts to put a man into orbit, onto the moon, and eventually, to explore space.

***Sputnik*:** The Soviet satellite that launched the American space race in late 1957. The Russians successfully launched *Sputnik* into orbit, signaling to the United States that they needed to excel in the sciences to regain a position of global superiority.

Critical Response

"Exploring the intimate relationships among blackness, womanhood, and 20th-century American technological development, Shetterly crafts a narrative that is crucial to understanding subsequent movements for civil rights." —*Publishers Weekly*

"The stories are amazing not because the women were extremely smart, but because they fought for and won recognition and devotedly supported each other's work. Their work outside the office—as Scout leaders, public speakers, and leaders of seminars to promote

SUMMARY AND ANALYSIS

science and engineering—was even more impressive."
—Kirkus Reviews

"Much as Tom Wolfe did in The Right Stuff, Shetterly moves gracefully between the women's lives and the broader sweep of history ... Shetterly, who grew up in Hampton, blends impressive research with an enormous amount of heart in telling these stories."
—Boston Globe

About Margot Lee Shetterly

Margot Lee Shetterly grew up in Hampton, Virginia, and her father worked for NASA. Her daily life was filled with the engineers and mathematicians behind some of the most audacious accomplishments in the US flight and space programs. After graduating from the University of Virginia, she moved to New York and worked in investment banking. In 2005, she moved with her husband, writer Aran Shetterly, to Mexico and founded an English language magazine dedicated to expatriates, in addition to working as marketing and editorial consultants for the Mexican

SUMMARY AND ANALYSIS

tourism industry. Currently, she lives in Charlottesville, Virginia. Through her family and personal relationships, Shetterly had direct access to the relatives of the women who changed history in the early days of America's aeronautics and aerospace technology programs, and their insights to life during the tumult of segregation and integration of Jim Crow America.

Shetterly is the founder of the Human Computer Project which seeks to tell the stories of the female scientists, mathematicians, and engineers who strengthened both NACA's and NASA's programs.

For Your Information

Online

"10 Unapologetic Books About Race in America." EarlyBirdBooks.com

"Ashes, Art, and Other Surprising Things Humans Have Left on the Moon." ThePortalist.com

"'*Hidden Figures*' author tells the story of the black women who helped win the space race." ESPN.com.

"'*Hidden Figures*': How Black Women Did The Math That Put Men On The Moon." NPR.org

"*Hidden Figures*: Meet the Black Female Math Geniuses Who Helped Win the Space Race." TheRoot.com

SUMMARY AND ANALYSIS

"*Hidden Figures* trailer: NASA's overlooked black female mathematicians." TheGuardian.com

"The Human Computer Project."

"On 'Hidden Figures' Set, NASA's Early Years Take Center Stage." Space.com.

"On Being a Black Female Math Whiz During the Space Race." NYTimes.com.

"Uncovering a Tale of Rocket Science, Race and the '60s." NYTimes.com.

Books

The Color Purple by Alice Walker

The Girls of Atomic City: The Untold Story of the Women Who Helped Win World War II by Denise Kiernan

Lab Girl by Hope Jahren

MLK: An American Legacy by David J. Garrow

Moon Shot by Deke Slayton, Alan Shepard, and Jay Barbree

Rise of the Rocket Girls: The Women Who Propelled Us, from Missiles to the Moon to Mars by Nathalia Holt

We Could Not Fail: The First African Americans in the Space Program by Richard Paul and Steven Moss

Women in Science: 50 Fearless Pioneers Who Changed the World by Rachel Ignotofsky

X-15 Diary: The Story of America's First Space Ship by Richard Tregaskis

Bibliography

Author Q&A: Margot Lee Shetterly reveals NASA's "Hidden Figures," September 11, 2016, http://www.collectspace.com/news/news-091116a-hidden-figures-shetterly-interview.html.

WORTH BOOKS
SMART SUMMARIES

So much to read, so little time?

Explore summaries of bestselling fiction and essential nonfiction books on a variety of subjects, including business, history, science, lifestyle, and much more.

**Visit the store at
www.ebookstore.worthbooks.com**

MORE SMART SUMMARIES
FROM WORTH BOOKS

POPULAR SCIENCE

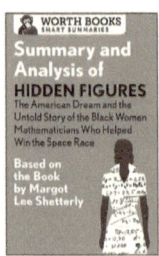

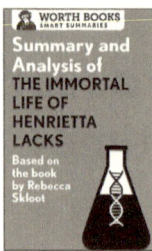

MORE SMART SUMMARIES
FROM WORTH BOOKS

EMPOWERMENT

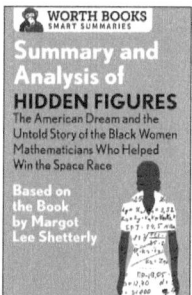

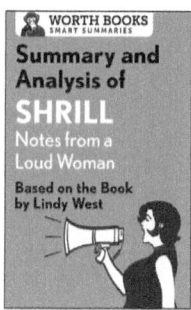

MORE SMART SUMMARIES
FROM WORTH BOOKS

HISTORY

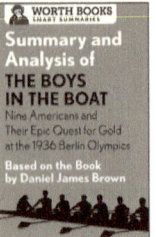
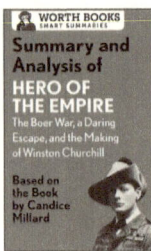

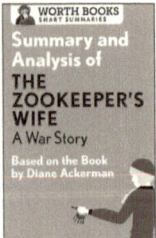

EARLY BIRD BOOKS
FRESH EBOOK DEALS, DELIVERED DAILY

LOVE TO READ?
LOVE GREAT SALES?

GET FANTASTIC DEALS ON BESTSELLING EBOOKS DELIVERED TO YOUR INBOX EVERY DAY!

INTEGRATED MEDIA

Find a full list of our authors and titles at www.openroadmedia.com

FOLLOW US
@OpenRoadMedia

www.ingramcontent.com/pod-product-compliance
Lightning Source LLC
Chambersburg PA
CBHW060342080526
44584CB00013B/885